John Croumbie Brown

African Fever and Culture of the Blue Gum-tree

To Counteract Malaria in Italy

John Croumbie Brown

African Fever and Culture of the Blue Gum-tree
To Counteract Malaria in Italy

ISBN/EAN: 9783744757904

Printed in Europe, USA, Canada, Australia, Japan

Cover: Foto ©berggeist007 / pixelio.de

More available books at **www.hansebooks.com**

AFRICAN FEVER

AND

CULTURE OF THE BLUE GUM-TREE

TO

COUNTERACT MALARIA IN ITALY.

COMPILED BY

JOHN CROUMBIE BROWN, LL.D., &c.

CONTENTS.

I. *Circular addressed to Secretaries of British and American Missions to Foreign Lands.*

Haddington, 25th December, 1889.

'Gentlemen,—I send herewith a copy of a circular which I issued last year offering a gift of treatises on Forestry free of charge to Librarians in British Colonies, and in the United States of America. There remain with me still several complete sets, and of some of the treatises a good many. If it seem to you as it does to me after a lifelong interest in Missions to the Heathen, intercourse with Missionaries to various lands, and prolonged and repeated co-operation with them, first in Russia, and subsequently in South Africa, that it might be useful to missionaries at some stations to have at command information on some of the subjects treated of in these volumes, I shall with pleasure place at your service one, two, three, or more sets complete or partial as may be desired, or as the copies on hand may permit, forwarding them in the order in which applications may reach me.'

The following is an excerpt from the circular referred to :—

In a Volume which I published last year under the title of *Schools of Forestry in Germany, with Addenda Relative to a Desiderated British National School of Forestry*, I stated :—

'While a British National School of Forestry might be made available for instruction in Modern Forestry to any party who may desire this through the medium of the English language, I know of no insuperable difficulties in the way of such institutions being organised in any of the States of the American Union, or in any of the Colonies of the British Empire, in connection with existing educational arrangements of their own. As a contribution of information, which might be of use to any residents in these, in determining what might be done in the matter, I am prepared to deliver

4

free, to any address in Edinburgh or London, which may be given to
me, a copy, in sheets, of any or all of the following works, to be
placed in a Free Public Library in any of these States or Colonies,
on an application to me certified by the Government of the State or
Colony.

1. Origin and History of Schools of Forestry in Germany, with
Addenda Relating to the Desiderated School of Forestry
in Britain.
2. The School of Forest Engineers in Spain, indicative of a
Type for a British School of Forestry.
3. Introduction to the Study of Modern Forest Economy.
4. French Forest Ordinance of 1669, with Historical Sketch of
Previous Treatment of Forests in France.
5. The Forests of England, and the Management of them in
By-gone Times.
6. Forestry in Norway.
7. Finland—its Forests, and Forest Management.
8. Forestry and Forest Lands in Northern Russia.
9. Forestry in the Mining Districts of the Ural Mountains in
Eastern Russia.
10. Forestry in Poland, Lithuania, and the Baltic Provinces of
Russia.
11. Pine Plantations on Sand Wastes in France.
12. Reboisement in France ; or Records of the Re-planting of the
Alps, the Cevennes, and the Pyrenees, with Trees,
Herbage, and Bush, with a view to Arresting and Pre-
venting the Destructive Consequences of Torrents.
13. Hydrology of South Africa ; or Details of the former Hydro-
graphic Conditions of Cape of Good Hope, and of Causes
of its Present Aridity, with Suggestions of Appropriate
Remedies for this Aridity.
14. Water Supply of South Africa, and Facilities for the Storage
of it.
15. Forests and Moisture ; or Effects of Forests on Humidity of
Climate.'

Of that offer the Directors of above a hundred Libraries in the
United States, British Colonies, Great Britain, and on the Continent
of Europe, have availed themselves. But the offer was limited to
Free Libraries ; and this has repeatedly been brought under my
consideration. To meet difficulties which I have been felt, I remove
the restrictions, and proffer the gift to any one or more of three
Libraries in each Colony or State, irrespective of their being Proprie-
tory, Subscription, or Free, on the condition of their meeting the
expense of transmission from Edinburgh, the place of Publication.
Applications will be met in the order in which they may be received ;
and should it happen that more than three applications be received

from any one Colony or State, the others will be met to the extent practicable from the Volumes at command, at the expiry of Six Months from this date. By arrangements of the Smithsonian Institution, Washington, gifts of Books can be transmitted free of charge from London to any Town in the United States. The offer is made of the Books in Sheets, partly because this is preferred by the Managers of many Public Libraries ; but they may be bound here at an average charge of 8d. per Volume.'

'At the risk of appearing to offer with one hand, and to beg with the other, which would not correctly represent my design, I may state that times innumerable I have been grieved at the loss of valuable lives through Malarious Fever in Central Africa, and at some Missionary Stations in other lands. I have not the knowledge required for a comparison of these fevers with the Malarious Fevers of Italy, but I believe them to be similar if not identical; and under this impression I have prepared detailed accounts of what has been done, and what has been effected by Plantations of the *Eucalyptus Globulus* or *Blue Gum-tree* in counteracting the insalubrious effects of residences near marsh lands in that country, and of the natural history of the Microbe to which the fatal disease is attributed, and I am ready to give the manuscript to any one who will meet the expense of printing it, for extensive distribution in lands in which such fevers prevail, on the sole condition of fifty or a hundred copies being placed at my own disposal.

'The expense I reckon may be about £50. There are embodied in the memoir or report translations of two French reports in regard to the success of the enterprise, the expense of printing either of which might be about £5, and I am ready to carry either through the press on the same terms if the whole be not printed,

I am, Gentlemen,

Your fellow workman,

JOHN C. BROWN.

II. *Fatal Effects of Malaria in Equinoctial Africa.*

1. It is at times saddening to read details of ravages in the mission field occasioned by African fever, a type of fever by no means confined to that Continent. Some of the latest sketches of desolation which has thus been wrought occur in the narratives given by Professor Drummond of his exploratory journeyings between the East Coast and Lake Tanganyeka. In these at one place he writes: 'It was a brilliant summer morning when the 'Ilala' steamed into Lake Nyassa; and in a few hours we were anchored in the little bay at Livingstonia. My first impressions of this famous mission station certainly will never be forgotten. Magnificent mountains of granite, green to the summit with forests, encircled it, and on the silver sand of a still smaller bay stood the small row of trim white cottages. A neat path through a small garden led up to the settlement, and I approached the largest house and entered it. It was the Livingstonia manse— the head missionary's house. It was spotlessly clean; English furniture was in the room, a medicine chest, and familiar-looking dishes were in the cupboards, books were lying about, but there was no missionary. I went to the next house—it was the School, the benches were there and the blackboard, but there were no scholars and no teacher. I passed to the next it was the Blacksmith's Shop; there were the tools and the anvil, but there was no Blacksmith, and so on to the next and the next, all in perfect order and all empty. Then a native approached and led me a few yards into the forest, and there among the mimosa trees under a huge granite mountain were four or five graves; these were the missionaries.'

Take another sketch: 'We struck across a low neck of land, and after a few hours found ourselves suddenly on the banks of the Zambesi. I saw that a red bungalow was in sight, and opposite it the little steamer of the

Africa Lakes Company, which was to take us up the
Shiré. There is more in the association perhaps than in
the landscape to strike one as he furrows the waters of
this virgin river. We are fifty miles from its mouth;
the mile-wide water is shallow and brown, the low sandy
banks fringed with alligators and wild birds; the great
deltoid plain yellow with sun-tanned reeds and sparsely
covered with trees stretching on every side; the sun is
blistering hot; the sky as it will be for months, a mono-
tonous d me of blue—not a frank, bright blue like a
Canadian sky, but a veiled blue, a suspicious and Malarious
blue, partly due to the perpetual heat haze, and partly to
the imagination, for the Zambesi is no friend to the
European, and the whole region is heavy with saddening
memories.

'This impression perhaps was heightened by the fact
that we were to spend that night within a few yards of
the place where Mrs. Livingstone died. Late in the
afternoon we reached the spot—a low ruined hut a
hundred yards from the river's bank, with a broad
verandah shading its crumbling walls. A grass-grown
path straggled to the doorway, and the fresh print of a
hippopotamus told how neglected this spot is now.
Pushing the door open we found ourselves in a long dark
room, its mud floor broken into fragments, and remains
of native fires betraying its latest occupants. Turning
to the right we entered a smaller chamber, the walls bare
and stained with two glassless windows facing the river.
The evening sun setting over the Morumballa mountains
silvering with his soft glow, took us back to that Sunday
twenty years ago, when in the same bedroom at this
same hour Livingstone knelt over his dying wife, and
witnessed the great sunset of his life. Under a huge
Baobab tree—a miracle of vegetable vitality and luxuri-
ance—stands Mrs. Livingstone's grave. The picture in
Livingstone's book represents the place as well kept and
surrounded with neatly planted trees; but now it is an
utter wilderness marked by jungle grass, and trodden by
beasts of the forest; and as I looked at the forsaken mound

and contrasted it with her husband's tomb in Westminster
Abbey, I thought perhaps ·the woman's love which
brought her to a spot like this, might be not less worthy
of immortality.'

Take yet another sketch—' Our next stoppage was
to pay another homage—truly this is a tragic region—at
another white man's grave. A few years ago, Bishop
M'Kenzie and some other missionaries were sent to Africa
by the English Universities, with instructions to try to
establish a mission in the footsteps of Livingstone. They
came here ; the climate overpowered them ; one by one
they sickened and died. With the death of the Bishop
himself the site was abandoned, and the few survivors
returned home. Among the hippopotamus trampled
reeds on the banks of the Shire, under the rough iron
cross, lies the first of three brave Bishops who have
already made their graves in Equatorial Africa.'

If we go again to the sacred spot, the grave of Mrs.
Livingstone, we find that there have been deposited there
the remains of yet another victim.

In the Report of the committee for the propagation of
the gospel in foreign parts to the General Assembly of
the Church of Scotland in 1888, in the account of
missionaries who had sailed in the course of the previous
year for East Africa, it is stated—' The first and larger
missionary party reached Quilimane on 25th April.
Before leaving this country they were assured that the
new steamer of the African Lakes Company would be
ready for them on their arrival. It was not ready ; and
after a detention of ten days they proceeded up the
Kwakwa, to make the ascent of the Zambesi, in the old
steamer, the Lady Nyassa. But the old steamer quickly
broke down, and they were stranded for weeks at
Vicenti, on the Zambesi, near the confluence of the Shiré
with the Zambesi, in a low and unhealthy country.
' Mrs. M'Ilwain had been married only two months

before at Braco, Perthshire, where her parents, Mr. and
Mrs. Cameron live. She was a younger sister of Mrs.
Duncan, who died at Blantyre a few years ago. Several
of the party suffered from malarious illness; but with
Mrs. M'Ilwain the fever was very intense, and ran its
course in four days. She died on her twenty-first birthday, Sunday, the 29th of May, the day on which, as
already mentioned, we here were ordaining another
missionary for Africa [the Rev. Robert Cleland], who was
on that day, during the sitting of the Assembly, ordained
and commended to God as a missionary about to proceed
thither. She was buried at Shupanga, close to the
grave of Mrs. Livingstone. David Livingstone wrote in
his *Last Journals*—'Poor Mary lies on Shupanga Brae.'
And doubly sacred is now the place where this young wife
lies, as truly given for Africa as if she had spent a
lifetime in the mission field.

In the *Mission Record* of the Church of Scotland for
September, 1887, appears a letter of which the following
is a copy:

Vicenti River, Zambesi,
30th May, 1887.

My dear Mr. M'Murtie, — I have very very sad news. Mrs.
M'Ilwain died here yesterday of fever of a very intense kind. Could
Dr. Rankin break the news to her poor mother and father, for whom I
am very sorry? I know what the grief in parting with her was to
them, for Mrs. Scott and I were at her wedding at Braco a few days
before we all left for Africa; this is the second daughter given for
the kingdom of Christ in the heart of this dark land. Her death
was somewhat like her sister's in the heroic noble hold she had of
life and death. And she died conscience of entrance into heaven.

The fever ran its course in four days. I thought she was gone
in the first rush. Her pulse rose to 106 and beyond. At first she
said she could not see Christ and wished to see Him; and we prayed,
and she did see Him. It was only for our sakes she consented to
pray that she might come back to life: when she did come the strong
life instincts returned, and she did do all she could. We got the
fever under before the second rise of temperature and the heart kept
steady: the third rise we also kept under, though it continued longer
than the second; but the fever seemed in the fourth rise to gather
frightful force, and then the heart failed too. It (the temperature)
went up rapidly to 106, 107, 108, and past, and then we measured no

more. The remedies seemed each to do what we wanted, and yet nothing could stay it. She was almost gone, then she rallied, and then the breathing ceased as Miss Beck was holding her head on her knee. It was her twenty-first birthday, Sunday the 29th May.

'Poor Mr. M'Ilwain is down at the month of the river ; I don't know what this will be to him. He is a noble fellow. Both he and Tanner are there helping on the steamer ; and we are here battling. She must be buried to-day ; and we have selected Shupanga, where Mrs. Livingstone is buried, as the spot. Here the bank washes away every year. The grave is being dug, and Mr. Henderson of Mompea is making the coffin. I can't tell you what this is to us all. One thing we have all learnt—to give ourselves more, body and soul for the work—not for ourselves, not for our Mission, not even to one another—all that is ours—but to God's work among these people in this seven-barred land. You've no idea how death brings down the bridge behind a mission party and launches us into God.

DAVID CLEMENT SCOTT.

The following is an extract from an account of the death of Mrs. M'Ilwain in *Life and Work*, September, 1877, given by the Rev. Dr. Scott, St. George's Church, Edinburgh :— 'Her soul is with the saints, this we may safely trust, while her earthly remains have been laid side by side with those of Mrs. Livingstone, so that those who love her and mourn her have the solace that her body in the lone wilderness of Shupanga has been gathered to very honoured dust. The Lord in his tenderness will be their comforter.'

It is sad ! Mention is made of Miss Beck, a lady of whom it is stated in the Report laid before the General Assembly, that she takes no salary—her sisters at home providing what is necessary by their own industry. In a letter to her sisters at home, she wrote of her :—'She took fever on Thursday morning, and before night her pulse had risen to 168, and temperature to 105·8. We got that under, and also the second attack next day ; there was a third attack on the following day, not so bad, but from the beginning of the fourth attack, on the fourth day, when her temperature rose to 108, we knew there was little hope. She died on the fourth day. The doctor who afterwards arrived seems to think that nothing would have saved her as it must have been a malignant fever from the first; and that the system must have been poisoned by a great amount of malaria.'

And of herself she writes:—' On Sabbath morning I did not feel very well at breakfast, the result of having been up late on the previous night. I intended to go back to bed, but as it was an exciting day did not manage this till about two o'clock, when I drank some tea to induce perspiration, and lay down. In a short time they found that my temperature was up to 105 and pulse 150. I did not feel very ill nor at all alarmed, as I had often felt as ill during an attack of sore throat. There was no pain and I was only in bed for two days. The fever returned very frequently afterwards for months, leaving me very well in the interval. It gradually diminished in frequency till there were about three weeks between the attacks.'

She wrote thus:—But I was told so ill was she on the day of the funeral that the little company bore away the remains of the departed under an impression that on their return they would have to proceed with arrangements for the burial of the writer. What was written was written afterwards: she was then unconscious or delirious, and wrote from what the friends who were well had told her. In the same letter she wrote: 'Mr. M., another member of the mission, is down this forenoon with a sharp attack of fever. He was at breakfast this morning, and may be up to-morrow again. I have not had any fever for three or four days, and am feeling all right.'

Two months later she wrote:—'14th July, 1887.—I have had fever thrice since coming here. Dr. B. thinks I may have it now and again for a little while yet. One may have fever at night and be up in the morning. 25th July.—But having fever occasionally takes up much more time.'

Happily she recovered, and some eighteen months later she wrote under date of 14th December, 1888:—'I had a little, a very little fever that night. It returned

for three nights, and I am better now. I had not had
fever for more than a year before.'

But while she thus recovered at Blantyre, a younger
patient, with more than youth in her favour, had died.
Mr. Hetherwick, in a very interesting Mission Journal,
writes of the death of a native girl at Blantyre:

Monday, 25th April.—Several children have come from Chirazula
to stay with us. This is the first time we have had any from such a
distance. Saturday, 21st May.—This afternoon one of the manse-
house girls died after twenty-four hours' illness. She took ill
yesterday afternoon with symptoms of severe fever. At evening she
was a little better, but this morning dangerous symptoms manifested
themselves, and she sank into a comatose condition. Everything
possible was done for her, but to no purpose, and at two o'clock she
died. She was the first of the Chirazula girls who came to Blantyre,
and I am anxious as to how the people may look on the matter.
Sunday, 22nd May.—The parents and friends of the girl arrived
. . . Very touching was the stifled grief—stifled for our sakes, because they knew we should not wish the usual native wailing.
Monday, 23rd May.—The girl who died on Saturday was buried to-
day . . . I spoke by the open grave of the resurrection from the
dead ; but how could they comprehend it on whose ears the news had
fallen for the first time ? If anything could make a man a missionary,
'tis the sight of a native mourning and funeral. Is it not something
to be even the bearer of the tidings of Him who said, 'I am the
Resurrection and the Life ?' Tuesday, 24th May.—The Chirazula
people left to-day—the mother saying she would be back soon : 'her
heart was here now where her child had died.'

As a companion picture to those given by Professor
Drummond of this region may be given the following by
Captain Lugard from an interesting article entitled—'A
glimpse of Lake Nyassa,' which appeared in *Blackwood's
Magazine* of January last. '. . . Hurriedly we pass
over the four or five days' boat journey up the Kwakwa
river from our starting point on the coast at Quilimane,
for this is not 'a thing to memory dear.' Yet it has its
fascinations. The trees on either bank drooping into the
silent river, great-limbed acacias with their feathery
foliage and sweet-scented yellow blossoms, and large fig
trees festooned with giant creepers ; the glorious, though
for you too powerful, sunshine ; the cheery and incessant
boatmen's song as they dip their paddles simultaneously

to the cadence ; the lov.ly tinted reed-warblers and gay butterflies ; and soaring above all the great fish-eagle with his weird cry, more weird and striking even than the lion's roar. All these are new to us, and charm us. But as evening falls the curse of the Kwakwa, the myriads on myriads of mosquitoes come forth—a misty army athirst for blood—and we are only too glad to light a fire of green wood and sit in the blinding smoke, though it draws involuntary tears from our eyes, to escape the smarting pain of their venemous bites. The night is a weary one, for no mosquito-net seems proof on the Kwakwa ; and in our very dreams we wonder what human fiends the Buddist creed deems to be fit tenants after death of the mosquito world. The indispensible morning dip is somewhat hazardous, for the river is full of crocodiles. So we reach the end of the Kwakwa journey where we must transfer ourselves and our effects across to the Zambesi, a portage of four miles. This is a noble river, perhaps three-quarters of a mile wide here. We are still in the fever zone, and shall be till we leave the river. Almost opposite is the mighty baobab tree, under which lies Livingstone's wife, and other graves around testify to the fatal malaria of the place. . . . Soon we branch off to our right, and ascend the Shiré, a broad, fine river, running into the Zambesi, and push on through the Morambala Marshes, with the great mountain looming in front and on our right ; and if you have not already had an attack of fever I can promise it to you here. As we wind in and out along the dead flat country the Morambala mountain assumes new shapes, and is now on our right and now on our left, and anon almost behind us, so interminable are the windings of the river, but it seems to get no nearer.

' Islands of water-cabbage and vegetable debris of all kinds float past us, and so we reach a pass through the Elephant marshes. Doubtless in no very distant past all this country was a shallow lake over which the great Morambala kept silent watch in the times when these dreary swamps had never echoed to the sound of the

white man's rifle. Many indeed believe that the ancient
Portuguese discoveries of a great lake with a mountain
rising out of it referred to these swamps then flooded, and
not to the real Nyassa as now alleged. . . . Next day
we are at the end of our first river voyage. A weary
climb for thirty miles up gradients which highlanders
may think trifling—for we are among Scotsmen now—
but which an average man is glad to see behind and below him, and we are at Mandala and Blantyre, some 3000
feet above the level of the sea. We have agreed to
journey on the lake, but you must halt here, or you
would outrage the generous Scotch hospitality; besides
there is only one Blantyre in Africa, and nothing like
it anywhere else. Savage Africa lies all around, but
passing up the long avenue of blue Eucalypti we find
ourselves in an oasis of civilisation, the more striking and
complete from the contrast. Well built and neatly
thatched houses of solid brick enclosing a square beauti-
fully kept in shrubs and flowers, all watered by a highly
skilful system of irrigation channels—which bring the
water from a distant brook—give a British homely charm
to the picture, and disarm surprise when we find well
stocked kitchen gardens, carpenters' shops, brickmaking
and laundry establishments all around us. The mission
children are dressed in spotlessly clean clothes, and look
bright and happy. . . . The children are not hap-
hazard comers, here to-day and absent by some whim to-
morrow, but boarders, many coming from far, the sons of
chiefs and headsmen. Over this little model colony
presides the *Genii Loci*—Rev. D. C. Scott, and his wife;
and I know not which exercises the greatest influence for
good. This influence is extraordinary for no one more
quickly recognises the real gentleman than the African
savage. It is a tempting spot to linger in, either in fact
or on paper. I would like to write fully of the Shiré
Highlands, of the very pretty church, so pretentious in
its architectural beauty as to have gained the sobriquet
of the 'Blantyre Cathedral;' of Dr. Bowie and his work;
of Mandala, and of Mr. Moir's many experiments with

dogs, goats, horses, poultry; or of the coffee plantations, and of Mr. Buchanan's sugar and coffee at Zomba; but we must push on to Nyassa, and I must refer you to Mr. Buchanan's interesting book.'*

Thus far Captain Lugard—but I must pause to remark that here it was, in the midst of such surroundings, that the young pupil at the Mission School died of fever; that here we read of a beginning made in the culture of the Eucalyptus; that here Miss Beck resided for years after her attack of fever on the river; and that possibly the little girl had brought the disease with her in an incipient form. I know nothing more than has been stated; and this is by far too little to warrant me to press any argument which might be raised on such facts; but with the magnitude of the interests at stake in view, I do not feel free to pass them without remark. Blantyre is a missionary station of the Church of Scotland.

Of Livingstonia, Captain Lugard writes:—'The view as we steam in the little 'Ilala' round Cape Maclear into the old mission of Livingstonia is picturesque in the extreme. A few feet from the shore we are in blue water, many fathoms deep, yet so clear that we can see the bottom as it were but a foot below us, and the water-growths, and shells, and many coloured fishes. Above and behind us tower the rugged boulders of the hills, and in front of this fairy port are a series of islands standing out to sea, behind which the setting sun shows with a glory that is superb as it sinks below Nyassa Lake. Much as I have travelled I have seen I think no lovelier spot in my life. Clear as crystal to look at, the water of Nyassa proves under analysis to be as good as it looks. There is singularly little flotsam cast ashore by the waves, and no floating debris on the waters, spite of the large quantity of wood and vegetation borne into the lake by the rivers then in flood.' Where could a more delectable site for a home be found? Yet this is Livingstonia which had been ravaged by the fever!

* "The Shire Highlands " by J. Buchanan.

Of the new mission of the Free Church of Scotland at Bandawé, he writes:—'Skirting up the west coast we come to the mission station of Bandawé on the lake shore, lat. 12° Dr. and Mrs. Laws have effected wonders here; their schools are thronged and the practical nature of the work is invaluable. But I must not allow myself to digress into a description of an African mission station, however tempting. Dr. Laws' contributions to science and his extensive information have made his name celebrated as the scientific referee on all Nyassa 'ologies. The people among whom he lives and works are the Atonga.'

Dr. Laws was connected with the United Presbyterian Church in Scotland; but was given by himself, or by the mission board of that Church, or by both conjointly to the Free Church Mission in this region, as Miss Beck, a member of the Free Church, gave herself at the desire of Mr. Scott, to the mission of the Church of Scotland at Blantyre. Thus are different sections of the Church co-operating to elevate the natives of the land in knowledge and civilisation, in godliness and virtue.

2. Of scenery on the Congo, like graphic and pleasing pictures, have been given; but there too fever has been in like manner destructive of life amongst those who have gone thither to carry on the work of Christ.

Even since this year began, intelligence has reached us that there another victim has fallen, 'perished in all the leaves of his spring.' *The Missionary Herald* of the Baptist Missionary Society, issued in February last, tells of the arrival at Underhill Station of four missionaries from that Society to the Congo tribes, one a missionary returning to his self-denying work, the others entering for the first time upon the field. In the next issue, that for March, appears an engraving from a photograph of them, taken at Madeira, by a friend of the mission, with a sister in the field. But the same issue tells that one of them was now no more a suffering

soldier of the Cross; ere the announcement of his arrival
had reached this land, he had been called to join the
great multitude of every nation and kindred and tongue
and people who have washed their robes and made them
white in the Blood of the Lamb, and now stand before
the throne of God. Within a fortnight after his arrival
he died. Mr. Oram, the missionary at San Salvador, the
station at which the youthful missionary died, thus
communicates to Dr. Underhill the circumstances of
his death:—

San Salvador,
December 31st, 1889.

MY DEAR DR. UNDERHILL,—It grieves me much that my first
letter to you should have to be a report of such sad news as I have
now to write.

Our brother, William F. Wilkinson, arrived here on the 14th
of this month to assist during the absence of Mr. Lewis; and it was
hoped that on the return to the station of Mr. and Mrs. Lewis, and
Mr. and Mrs. Graham, he and I would have the pleasure of travelling
to Wathen together.

For the first week our brother had capital health, and he was
rejoicing to think that he had escaped fevers for so long a time. He
had his first experience of African fever when I myself had a slight
touch of it, which kept me in bed on the 23rd inst., during which
day he conducted the business of the station. The following
morning I was up to breakfast, but Mr. Wilkinson complained of
headache and a poor night, so after a slight breakfast he returned for a
rest to his bedroom. At noon, I went in to take his temperature, and
found he was in fever. He was rolled up in blankets, and perspired,
but the temperature could not be brought low enough to take quinine
effectually till the following morning, when I administered about 20
grains while a chance was offered. It was well that I did so, for his
temperature did not decline below 103 during the rest of this and
the following day.

He was watched carefully by senior lads, besides my own
attendance, who did for him all that a sick man requires. Nlekai,
our assistant schoolmaster, stayed awake with him on alternate
nights, but would frequently come in during the day also.

On Saturday, the fever seemed to abate somewhat, and then he,
for the first time, felt the quinine at work by the usual humming in
the ears. We all entertained great hopes for the Sunday, especially
as in the evening he felt a desire to sleep. I had just gone to bed,
when the boy who was with him came in to tell me that Mr.
Wilkinson had seen a snake on his bed. I went in to find him in
a great state of alarm, and out of bed. I searched, but, as I

expected, found no snake ; he then told me of strange dreams he had dreamed. He became quieter and I returned to my bed (because for my own health I knew I must not stay up night and day), leaving him with his attendant. At midnight Mr. Wilkinson awoke again and sent in for me. He told me of other dreams, which showed that his mind was wandering. I then sat with him a while and read some favourite hymns for him to fix his thoughts upon ; he then repeated one of his favourites.

Before leaving I sent for Nlekai, who remained with him through the night. The Portuguese doctor was in attendance.

But in spite of all, the temperature continued to rise and the delirium accordingly with extreme violence. Once he spoke of his desire to be engaged in the work at Bolobo, and his sorrow that now he could not go there ; at another time he offered aloud a long earnest prayer for the work, quite intelligible although it was in delirium. But later on the mind became confused. The doctor and an English-speaking gentleman came in three or four times and showed very great sympathy and kindness, but when the fever mounted up to 106, and finally to 107·8, the doctor saw it was a bad case.

At 9·30 delirium suddenly ceased, coma set in, and the breathing became harder and more gasping, and finally ceased at ten o'clock.

The end was a wonderful contrast to the wild delirium of the whole Sunday. The boys were all present when they knew that the end was nigh, and over the bed we joined in prayer before we parted for the night. But though we went to bed, there was little sleep for some of us. To say I am sorry he is gone is saying little. I miss him intensely. For fifteen days only we were together. He was very earnest in seeking souls, and in the work he undertook found much delight. The blank his death has made causes much loneliness. Mr. and Mrs. Lewis will, I trust, be here by the end of next week, as also Mr. and Mrs. Graham, and probably they will write to you further and also to the friends of Mr. Wilkinson. A very full account of the fever has been recorded, which by next mail shall be sent for Dr. Roberts' inspection.

The chapel was filled with Congo friends at the funeral yesterday. The Resident and Commandant, also the Dutch and Portuguese traders (who displayed the warmest sympathy), were present at the grave.

With kindest regards to yourself and deepest sympathy with the bereaved friends at home, and not least with the work which calls for men and help,

I am, my dear Dr. Underhill,

Yours very faithfully,

FREDERIC R. ORAM.

I have no reason to suppose that the missionaries of the

Society by which Mr. Wilkinson was sent to the Congo have been exceptionally exposed to the malaria of the region to which they went. Some of them were exceptionally favourably prepared for the occupation of such a sphere of labour, having been in some measure acclimitised by previous residence and labour at the Cameroons: and yet, from a list of the missionaries' reports, it appears that of forty-two missionaries, inclusive of ladies, sent out in the decade, 1877-1886, there were only sixteen in the field in September last, 1889: one, an engineer, was in Britain, studying medicine; another, a lady, was in Britain, learning nursing—both with a view to future work. The following is the record of the rest :—

Rev. T. J. Comber, sent to Congo, 1877, transferred from Cameroons Mission, died on board s.s. 'Lula Bohlen,' and was buried at Mayumba, 27th June, 1887.

Mrs. T. J. Comber, sent to Congo, 1879, died at San Salvador, 24th August, 1879.

Rev. H. E. Crudgington, sent to Congo, 1879, invalided and sent to India, 1885.

Mrs. Crudgington, sent to Congo, ——, invalided and sent to India, 1885.

Rev. J. S. Hartland, sent to Congo, 1879, Died at Bayneston, 12th May, 1883.

Rev. H. Dixon, sent to Congo, 1881, invalided and sent to China.

Rev. W. H. Doke, sent to Congo, 1882, died at Underhill, 14th February, 1883.

Rev. H. W. Butcher, sent to Congo, 1882, died at Manyanga, 15th October, 1883.

Rev. W. Hughes, sent to Congo, 1882, invalided and at present in England.

Rev. S. A. Comber, M.R., sent to Congo, 1883, died at Wathen, 24th December, 1884.

Rev. Wm. Ross, sent to Congo, 1883, invalided and at present in England.

Rev. H. G. Whitley, sent to Congo, 1883, died at Lukungu, 3rd August, 1887.

Rev. J. W. Hartley, sent to Congo, 1883, died at Manyanga, 1st March, 1884.

Mr. Wm. Douglas, engineer, sent to Congo, 1883, died at Manyanga, March, 1884.

Mr. Arthur Baker, engineer, sent to Congo, 1883, died at Manyanga, March, 1884.

Mr. Mims, engineer, sent to Congo, 1883, died at Wathen, 27th
 September, 1884.
Rev. F. C. Darling, sent to Congo, 1884, died at Underhill, 19th
 March, 1887.
Rev. A. H. Cruickshank, sent to Congo, 1884, died at Wathen,
 27th March, 1885.
Rev. D. McMillan, sent to Congo, 1884, died at Underhill, 9th
 March, 1885.
Rev. A. Cowe, sent to Congo, 1885, died at San Salvador, 21st
 May, 1885.
Rev. W. F. Cottingham, sent to Congo, 1885, died at Underhill,
 8th June, 1885.
Rev. J. E. Biggs, sent to Congo, 1885, died at Arthington, 26th
 August, 1887.
Rev. J. Maynard, sent to Congo, 1885, died at Underhill, 28th
 January, 1886.
Rev. M. Richards, sent to Congo, 1885, died at Banana (mouth of
 Congo), 19th August, 1888.
Rev. S. Silvey, sent to Congo, 1886, transferred from Cameroons.
Dr. Seright, sent to Congo, 1886, invalided and returned to England,
 1886.
Rev. J. H. Shindler, sent to Congo, 1886, died at Underhill, 19th
 March, 1887.

 I have not the testimony which would warrant me to
say that all of these were the victims of Malarious Fever.
But I know that that fever is accompanied by so many
complications, is so frequently masked by other diseases,
and is so prevalent, of such frequent occurrence, and
leaves so many predispositions to the same or other
diseases — that I have no doubt that a majority,
comprising almost the whole of them, may reasonably be
reckoned victims of that malady: and, continuously,
additions are being made to the number of the slain.
The death of Mr. Wilkinson, who arrived on the Congo
in the middle of December, and was buried on the 30th of
the month, closes the record for 1889, and on the 2nd of
January, 1890, the record of another year was opened for
another victim to African fever on the west coast.

 3. Passing northwards to the Gulf of Guinea, the whole
southern coast from the Cameroons by the Bight of
Benin, the Gold Coast, and onward to Sierra Leone, we

21

pass through what has long been considered a hot-bed of
fever, though there also in the bright sunshine the
scenery may be considered enchanting—to the mortality
amongst missionaries to Calabar alone can I here direct
attention.

Thence Mr. Reeve Johnstone, a missionary there, in the
following letter to Mr. Oates, General Secretary of the
Young Men's Christian Associations and Fellowship Unions
of Scotland, reported the death of Mr. James Henderson,
who, with the writer of the letter, was well known to
many as one of the Christian workers in Glasgow before
going out to the mission field on the Calabar river:—

Mission House, O'Bombois Beach,
Umon, 6th January, 1890.

MY DEAR FRIEND,—Again it is my sad lot to be the medium of
mournful tidings—James Henderson fell asleep in Jesus on Thursday,
2nd January, at 7·45 p.m., at Ikoraua, in the house of Mr. Jarret,
the U.P. missionary there. His illness was of a bilious type, and
latterly a severe hiccough seemed to hasten the end.

His final illness lasted a week or so — he took ill on Christmas
day. I had been down myself with fever, and was getting better of
it when he took ague, which ran its course, when fever ensued, and
with it severe vomiting, which lasted with little intermission till his
death. I asked him if he would like to go to Calabar to see the
doctor. He said no. He said he would prefer going to Ikorana,
which is about ten miles up river from here, and stay with Mr. and
Mrs. Jarret. I had previously written to the Jarrets asking advice,
and they sent some medicine and an invitation to come up, so we
went up on Monday. It appears he got on well till Wednesday
morning, when the vomiting began again. He fought against it
manfully, until, feeling he could hold out no longer, he told Mrs.
Jarret he was dying. He seemed so bright and happy all the time
that Mrs. Jarret was slow to realise the true position. He said
'Yes, I am going. I came out to do the Lord's work, and it is but
little I have done; but if He sees it right to take me away, His will
be done.' Then tenderly referring to a young lady to whom he was
about to be married, he said, 'I will never see her on earth again;
but I will see her *in the morning.*' He bade the teacher's wife and all
around good-bye; then after Mr. Jarret had prayed with him, he
said, 'Amen,' and, composing himself for a little, he went smiling
into glory—the battle fought, the victory won.

Thus died James Henderson one of the sweetest deaths man
ever died, leaving a blessed testimony to the truth of the word,
'Though I walk through the valley of the shadow of death I will

fear no evil, for Thou are with me.' The valley was light, for Jesus was there. Mr. Jarret, who was present, says he thought he was falling asleep, when he fell into the arms of Jesus. Before he died he repeated several texts, and part of the hymn, 'Shall we all meet at Home in the morning?' Mr. Jarret sent for me post haste, but ere I arrived he was in heaven. On Friday, Mr. Jarret and I, with the teacher, made his coffin, and held a service at the house, to which a few came. Mr. Jarret read in I. Cor. xv., and ere he reached the end, broke down. He spoke a few words to those present, and with tears pleaded with them to turn to Jesus. At the grave he tried to sing, 'Shall we meet at Home in the morning?' but it fell flat; for who could sing with a sad heart? and, truly, mine was sad and heavy. A year ago I buried poor Gardner at Duketown, and now Henderson at Ikoraua. 'Who shall be next?' is the awfully solemn question which suggests itself. I am extremely sorry for Miss ——, who was just coming out to be married to our departed friend.

The United Presbyterian Missions to Calabar, like the missions of the Baptist Society to the Congo, has its honourable but saddening list of martyrs, men and not less noble women, who have hazarded their lives in the service of Christ, and have fallen 'counting not their lives dear unto them that they might finish their course with joy and the ministry they had received of the Lord Jesus to testify of the grace of God.' The Rev. Mr. Anderson, one of the oldest of the missionaries to Calabar, and at present in this country for an operation for Cataract, on my application has supplied to me the following list of missionaries to Calabar who are now dead, with the accompanying reference to volume and pages of the Churches' Missionary Record, in which notices of the several deaths are given.

Edmund Miller (Negro), Jamaica, at F. Po., Dec., 1846, see Record, May, 1847, p. 75.

Rev. Wm. Jameson, Scotland, Creek Town, Aug. 5, 1847, see Record, Jan., 1848, p. 4.

Alex. Sutherland, Scotland, Old Town, April, 20, 1856, see Record, Aug., 1856, p. 136.

Rev. Sam. Edgerley, sen., England, Duke Town, May 25, 1857, see Record, Aug., 1857, p. 148.

Mrs. Timson 1st, Scotland, Creek Town, Sept. 11, 1858, see Record, Jan., 1859, p. 9.

Henry Hamilton, Jamaica, Creek Town, Sep*. 23, 1858, see Record,
    Jan., 1859, p. 4.
Mary Stuart, Gala Water, Old Town, Sept. 26, 1858, see Record,
    Jan., 1859, p. 10.
Mary Cowan, Jamaica, Ikeropion, Sept. 13, 1862, see Record, Dec.,
    1862, p. 227.
Rev. Jo. Baillie, Scotland, Edinburgh, May 7, 1864, see Record, May.
    1864, p. 127.
Rev. Z. Baillie, Scotland, Liverpool, Aug. 4, 1865, see Record, Oct.,
    1865, p. 190.
Rev. Wm. Timson, Scotland, Ikamter, Jan. 10, 1870, see Record,
    Sept., 1870, p. 263.
D. E. Lewis, Wales, Creek Town, Aug. 8, 1870, see Record, Nov.,
    1870, p. 313.
Rev. Jo. Granger, Scotland, Ikamter, Dec. 9, 1870, see Record,
    March, 1871, p. 405,
Geo. Ashworth, England, Creek Town, Aug. 8, 1871, see Record,
    Oct., 1871, p. 644.
Euph. Johnston, Scotland, Creek Town, Jan. 5, 1873, see Report,
    Oct., 1873, p. 635.
Mrs. Timson 2nd (Margery Beety), Scotland, Duke Town, Sept. 20,
    1873, see Record, Dec., 1873, p. 685.
Alex. Morton, Scotland, Duke Town, Jan. 1, 1877, see Record, April,
    1879, p. 431.
Euph. Sutherland, Scotland, Duke Town, Oct. 14, 1881, see Record,
    Jan., 1882, p. 13.
L. Anderson, Jamaica, Duke Town, Jan. 26, 1882, see Record, April,
    1882, p. 92.
Rev. D. Williamson, Scotland, at sea, Jan. 30, 1882, see Record,
    March, 1882, p. 57.
Rev. S. H. Edgerley, Jamaica, Duke Town, Feb. 24, 1883, see Record,
    April, 1883, p. 153.
Mrs. Beedie 1st, Scotland, Scotland, Jan. 26, 1886, see Record, 1886.
Mrs. Beedie 2nd, Scotland, at sea, Oct., 1887.

On applying to Mr. Anderson for information in
regard to the immediate cause of death, he informed me
that he believed that Mr. Jameson was of the number of
those who died from fever; and that a goodly number
have died of diseases directly connected with the climate;
but that he can remember few cases of which he could say
that African fever was the immediate cause of death. It
is from what I know of the frequency with which
intermittent fever is masked by other diseases, and the
frequency with which death, though attributable directly

to fever, is the immediate consequence of complications thereby induced, that in the case of these, as in the case of those who have fallen in the Congo Mission of the Baptist Society, I have, after consideration, given all the names supplied to me.

Lately I had sent to me the following extract from a letter, from a missionary whom I highly esteem, to one of his former fellow-students: 'The damp—the perpetual drip—for it was the rainy season, tried the constitution rather severely. Again and again I went down under the prostrating touch of fever and arose but to fall once more.' And I have seen another letter of another date to another fellow-student, in which he says: 'It is of the utmost importance that this house be put up as early as possible, as to live in an African house in the rainy season is chiefly productive of fever: and I have had my share already of the embraces of this fiery foe.'

4. Of the bearers of evil tidings to Job we read that whilst the first was yet speaking there came a second; and while he was yet speaking there came also a third; and whilst he was yet speaking there came also another. When the circular prefixed to these details was issued, Mr. Wilkinson, a missionary to the Congo, was on his deathbed; and a few days thereafter he was in his grave. Scarcely was he buried when Mr. Henderson, a missionary to Calabar, sickened and died and was buried. And since these sheets were put into the hands of the printer, the Church Missionary Society has received a telegram from Zanzibar announcing the death from fever, of Mr. Alex. Mackay. When Mr. Stanley's memorable letter from Uganda appeared in Nov. 18, 1875, and led to the Church Missionary Society undertaking a mission to the Victoria Nyanza, Alexander Mackay was one of the first to offer his services. He was then a mechanical engineer at Berlin, and proved to be a singularly able and accomplished young man. He sailed for Africa with other members of the first party on April 27, 1876. On the journey inland he was taken ill, and was sent back to

the coast, but he refused to leave for England, and for
more than a year he was engaged in making a rough
road from the coast to Mpwapwa when news reached him
of the death of Lieutenant Shergold Smith and Mr. O'Neill
on the Victoria Nyanza. He pushed on to join the Rev.
C. T. Wilson, who was then alone in Uganda. He arrived
at Mteso's capital in December, 1878, and from that time
until July, 1887, Uganda was his home. At length the
bitter hostility of the Arab traders virtually drove him
out, but even then he would not come to England but
remained at the south end of the great lake, where Mr.
Stanley found him in September last, and where, it is
presumed, he has now died.

Under date of October 15, 1889, Mr. Stanley wrote
from Ugogo to Mr. A. L. Bruce, son-in-law of Dr.
Livingstone: 'I take this powerful body of native
Christians in the heart of Africa—who prefer exile for
the sake of their faith to serving a monarch indifferent or
hostile to their faith—as more substantial evidence of the
work of Mackay than any number of imposing structures
clustered together and called a mission station would be.
These native Africans have endured the most deadly
persecutions; the stake and the fire, the cord and the
club, the sharp knife and the rifle bullet have all been
tried to cause them to reject the teachings they have
absorbed. Staunch in their beliefs, firm in their
convictions, they have held together stoutly and
resolutely, and Mackay and Ashe may point to these
with a righteous pride as a result of their labours to the
good kindly people at home who trusted them. I
suppose you do not know Mackay personally. Well he is
a Scotsman—the toughest little fellow you could
conceive; young, too—probably 32 years or so, and bears
the climate splendidly—even his complexion is uninjured
—not Africanised yet by any means, despite twelve
years' continued residence. . . . Mackay plods on,
despite every disadvantage, sees his house gutted and his
flock scattered, and yet, with an awful fear of breach of
duty, clings with hopefulness to a good time coming,

when the natives of the country will be able to tell out
to each other the good news of 'peace and goodwill to
men'.'

Yes, another man of might has fallen!

The Editor of the *British Weekly* thus writes of
him:—

'The most popular missionary of the day is dead.' So the
Record last week commenced its article on Mackay of Uganda. It
seems but yesterday since he was my constant companion. He was
a native of Rhynie, the birthplace of James Macdonell, of the
Times, and the homes of the two families almost touched one
another. He was brought up under the shadow of Tap o' Noth, the
hill which I daily saw at a distance of but four miles. How well I
remember his bright boyish face, his blue eyes, his cheerful good-
nature, and his extraordinary mechanical ingenuity. He did not
look down upon his juniors, but was ever ready to join in schemes
of work and play. Considering that he was the son of one of the
most eminent geographers of the day, it is not wonderful that he
showed surpassing skill in map drawing ; some of his productions
are before my eyes at this moment. Cheerful, courageous, indefati-
gable, inexhaustible in resource, and devoted to the Kingdom of
Christ, he found his true sphere in Uganda ; and his name, as one
of the most illustrious and self-sacrificing pioneers of Christianity
and civilisation in Africa, will never be forgotten. The story of his
early years is told below by him who could most fitly tell it, and it
is needless to add a word :—

Mr. Mackay was the eldest son of the Rev. Dr. Mackay,
formerly Free Church minister at Rhynie, Aberdeenshire, now
residing at Ventnor, Isle of Wight. He was born in the Free
Church Manse, Rhynie, October 13th, 1849, and received his early
education at home, both parents contributing their share. Amid
his numerous other duties and engagements, pastoral and literary,
Dr. Mackay always made a point of devoting one hour daily to the
instruction of his children.

Short as this may appear to be, it is astonishing what great
progress may thus be made when the prescribed lessons have been
thoroughly prepared.

The future missionary evinced almost incredible precocity.
He could read the Bible at the early age of three ; and before he
was four he was well acquainted with the geography of Europe.
In his father's study there hung a large wall map of Europe by
Chambers. The little boy, taking a rod in his hand, would instantly
point to any country, capital, or other important town and large
river or mountain chain contained in the map. Before he was ten
years of age he had acquired a fair knowledge of all the branches

; to tell out
goodwill to

writes of

ad.' So the
Uganda. It
on. He was
nell, of the
touched one
o' Noth, the
How well I
eerful good-
He did not
a in schemes
f one of the
rful that he
productions
us, indefati-
Kingdom of
name, as one
Christianity
story of his
tell it, and it

Mr. Mackay,
nshire, now
in the Free
ed his early
are. Amid
and literary,
daily to the

what great
a have been

e precocity.
d before he
of Europe.
Europe by
ld instantly
a and large
he was ten
ae branches.

of a good English education. But one hour a day's tuition left only a very small margin for the elements of Latin and Greek, though these were by no means wholly neglected. Owing to local causes Dr. Mackay found it impossible to send his children to the parish school, very much to their loss.

Hence Alexander was sent at a very early age to the Grammar School of Old Aberdeen, in order to obtain a competent classical education, and especially to qualify him at the November term to carry a bursary or scholarship, so as to enable him to enter the University, where, during a curriculum of four years in the arts classes, his education would be without expense to his father, whose salary as a Free Church minister could not otherwise suffice to give a University education to his sons. For two years Alexander attended this very celebrated Grammar School, and added greatly to his stock of knowledge. But when the day of competition for bursaries arrived he found he had no chance amidst so many competitors, most of whom had thrown their whole mind into the art of translating English into idiomatic Latin. Indeed, it was this art that was mainly taught at the Grammar School of Old Aberdeen, for this alone would be of any service in carrying a scholarship in those days. His failure at the competition very grievously disappointed both of his parents, who had destined him for the ministry, should it be the will of God. In short, this was the turning point of his future career. Had he carried that scholarship, in all probability he would never have gone to Africa as a missionary.

But this could not be, for God had ordained that he should be instrumental in the conversion of so many precious souls in Uganda. Soon after this his mother died, at the early age of thirty-nine. He was the child of many prayers. She carefully instructed him in Divine truth, and early dedicated him to the Lord, whom she devoutly loved and faithfully served.

In 1867, Dr. Mackay resigned his ministerial charges in Aberdeenshire, and, with his family, went to reside in Edinburgh. It was determined that Alexander should study practical engineering, and he was sent to learn his future profession with Miller & Herbert, whose large establishment was in Leith Walk. Here, however, he was permitted to attend several classes at the University of Edinburgh. Amongst these were the higher Greek class, and the higher mathematical class, taught by Professor Kelland. In this class he distinguished himself, and at the end of the session, 1870, won a prize. He also studied engineering under Professor Jenkins, and received much valuable instruction by attending evening classes at the Mathematical Institution and School of Arts. Here he studied various sciences, and took a high prize in 1872. In 1875 he went to Berlin in order still further to perfect himself in his profession, and entered soon a large engineering house near it.

Early in 1876 his brother Charles, an ostensibly bright and promising boy of 15, died in Edinburgh, to the great grief of the whole family, and Alexander was much impressed by the event. It was about this time that he got introduced to Mr. Pearsall Smith, the well-known American evangelist, whose solemn and earnest entreaties induced Alexander to turn his back upon the world and become the bond-servant of Jesus Christ, and to resolve no longer to live to himself, but to Him who died for him.

A few days after taking this step he happened to glance at a *Times* newspaper, and observed an advertisement from the Church Missionary Society, Salisbury-square, wanting devoted young men to become their missionaries to Uganda in East Central Africa, where Mr. H. M. Stanley had induced Mtesa to agree to receive the heralds of salvation. He lost no time in answering the advertisement, and in offering himself to the Secretary. The Secretary expressed a wish to have a personal interview with him, and accordingly he set out for London without a day's delay.

He was readily accepted as a lay missionary and engineer. He agreed to go without any salary, but before formally engaging, he considered it to be his duty to consult his father. His father was more than delighted at what had taken place, and freely gave him over to the Lord and to Africa. Finding in Edinburgh his old and esteemed friend, Dr. John Smith, who was also a devoted Christian, he had little difficulty in inducing him to offer himself to the Society as their medical missionary. The party sailed from Southampton for Zanzibar, April 27, 1876.

Mr. Mackay's career thereafter is known to all interested in missionary work. Of the greatness and worth of his achievements and character the testimony of his close associate, the Rev. R. P. Ashe, is a sufficient testimony. The following may be quoted :—

It is impossible to do him any justice in some hurriedly-written lines ; but if one characteristic more than another made him a great missionary, it was his extraordinary patience and power of persisting in any work. He was never in a hurry, yet one work after another was taken in hand and finished. The amount of physical labour he would go through was astonishing. Nothing was a trouble to him, and he would not hear of the word impossible. Stanley, whose detractors allow his great power and discernment of character, on meeting Mackay called him the modern Livingstone. It was very high praise, but it was not too high praise, for he, like Livingstone, lived all for others and nothing for himself. Both were actuated by the strong and simple faith that God would surely and certainly save Africa, and both lived and died in order that this purpose might be accomplished.

Many of us had been looking forward to Mackay's return home for a well-earned rest. His name is a household word with all who love Africa or know anything of African Missions. But instead of this God Himself has laid him to sleep by the Nyanza.

Fred E. Wigram, Hon. Secretary of the Church Missionary Society, in announcing the death of this famous Scottish missionary, says :—

Mr. Mackay had a large share in the patient teaching of the people of Uganda, which resulted in the conversion of hundreds to Christianity, and in the reduction of the language to writing, with the translation of portions of Scripture, prayers, &c., and the preparation of 'reading sheets,' by which large numbers learnt to read. It was he who worked the little printing press which has supplied some thousands of copies of these fragments of literature for the instruction of the people. It was he who used his mechanical skill in house-building, boat-building, and frequent commissions of all sorts for the King of Uganda. He also contrived in a most surprising way to get out from England every kind of current literature, and to keep himself abreast of modern thought and progress. His letters and articles are those not only of the Christian missionary, but of the cultivated scholar and observant man of the world. It has not been sufficiently realised that it is to Mr. Mackay that we owe almost all the intelligence that reached England regarding Emin Pasha prior to Mr. Stanley's expedition. The first news that Emin was alive and holding his own was received by the same mail, in October, 1886, that brought, also from Mr. Mackay, the recovered last diary of Bishop Hannington : and Emin's letters and the bishop's dairy appeared in the *Times* on the same day. I am sure that the whole Christian world will agree that another of our truest African heroes is now lost to us, and I am sure that the succession will not be broken. The Rev. A. R. Tucker is to be consecrated on the 25th inst. as successor to Hannington and Parker in the bishopric of the Church of England in Eastern Equatorial Africa, and starts the same evening for East Africa. How many like-minded men will come forward and follow him ?

In the same issue of *The Scottish Leader*, in which this appeared, appeared the following :—

NEWS FROM CENTRAL AFRICA.—Dr. Moir, Castle Street, informs us that yesterday he received letters from Mandala, dated 26th February, but so far as the action of the Portuguese is concerned there is not much to tell. His son Frederick and his wife, who went out lately, were 18 days in reaching Mandala from Vicente, near the mouth of the Shire—it should have been done in six. This delay was greatly owing to the Portuguese having forbidden the natives to cut wood for British steamers, and threatening them with punishment if they did so. This therefore had to be done by the crew, necessitating constant stopping and delay. They had either scorching sun 102 degrees in the shade, or heavy rains, and the result was that both were laid down with fever, and were still under it when the mail left. Fred. was suffering most, but his older son was also suffering much, from fever induced by long-continued work and anxiety.

5. John tells—'1 heard a voice from heaven saying unto me write, blessed are the dead which die in the Lord from henceforth; yea, saith the spirit, that they may rest from their labours; and their works do follow them.' We bow the head and say, God is the will of the Lord! but it is with tear-filled eyes. Such is the stuff of which missionaries are made: and are such men so abundant that we can afford to let them be cut down one after another by fever—a fever which might apparently be prevented?

More than fifty years ago I was invited to accompany the secretary of the Education Society of New-England to different colleges in that quarter of America, to beat up for recruits for the mission field. His appeal generally began thus—We have at present openings for such and such extended operations in so and so, and in so and so. If any of you feel that with the openings before you at home, as merchants, professional men, and pastors of churches, you can with a free conscience remain, do so, and may God bless you in your work! You are not the men we want. But if there be any who with the call from their bretbren in the mission field feel that they cannot with an easy conscience remain at home whatever its prospects of life and labour and usefulness, you are the men we want. Such men have been found; and as one and another have fallen, one and another have advanced to take their place; but were it not better that the dead and the living had been this day labouring together? anJ wherefore not, if death from fever be a preventable death?'

And there are the representatives and agents of other missionary organisations who may have fallen in corresponding numbers, or in like proportions to the numbers who have gone forth to labour in the fields in which those have died. And there are many other fields of missions in Africa and

elsewhere in which the labourers are exposed to like
dangers. Nor are missionaries the only philanthropists
and enterprising emigrants who are seeking the good of
the benighted and the enslaved dwellers in that land,
who are exposed to the loss of life in the prosecution of
their enterprise. I write of missionaries because it is as
from a stand-point of observation within sight of the
mission field that I write; and it is to men who are the
supporters of Christian Missions that I am addressing
myself. But we know full well that explorers, and
merchants, and Government officials, and men and women
in their employment have also fallen, fallen in numbers,
and fallen perhaps unnoticed, though many of them,
noble-minded men and noble-minded women—Men and
Women well deserving a nation's homage. And these considerations add to the importance of everything practical
and expedient being done to secure the preservation of
useful lives.

The well-known editor of Livingstone's journals, the
Rev. Horace Waller, thus sums up his account of such
and of allied English enterprises in his *Title Deeds to
Nyassaland*. 'Dotted here and there from the Mangrove
Swamps of the Congo, south of the Zambesi to the
farthest extremity of Lake Nyassa, we pass the graves
of naval officers, of brave ladies, of a missionary bishop,
of clergymen, foreign office representatives, doctors,
scientific men, engineers and mechanics, all these were
our countrymen. They lie in glorious graves. Their
careers have been foundation stones, and already the
edifice rises.' With him and with others I take comfort
in the thought that the loss of life in these noble pioneers
has not been in vain. But earnestly do I desire that
such loss of life should be stayed.

In view of the aggregate number of those who have
fallen, in view of the sufferings they must have endured,
and of the extinction of their hopes—hopes of turning
many to righteousness; hopes of relieving the oppressed and
raising the fallen; hopes of providing for aged parents,

a widowed mother or an orphan sister ; the hope of a long and happy home life with some beloved loving companion, or hopes of securing a competence for life, or a fortune which would make old schoolmates stare, all extinguished like the snow-flake in the stream. In view of what has been and of what might have been, in view of the expenditure of care, of money, of fatigue, of thought, all to end in this; and in view of the sorrow and disappointments experienced by bereaved ones at home, one may cry aloud : O God, is there no remedy? Is it to go on thus for ever? and this not here alone, but also in many a land to which mundane considerations or higher aspirations like unto those of Him of whom it is written 'for the joy that was set before Him, He endured the cross despising the shame, and is now set down on the right hand of the Majesty on high,' may lead the sanguine, the enterprising the devoted? When prompted to cry aloud, Can nothing be done to prevent this loss of life? I think there can; and I feel as if some one of the bereaved were looking at me through tears of bitter sorrow, reminding me without words of the saying of Paul, 'To him who knoweth to do good, and doeth it not, to him it is sin;' and pointing me to the words of one who lived a thousand years before Paul, 'If thou forbear to deliver them that are drawn unto death, and those that are ready to be slain ; if thou sayest — Behold we knew it not, doth not He that pondereth the heart consider it? and He that keepeth thy soul, doth not He know it? and shall He not render to every man according to his deeds ?' But I cannot myself do what is requisite. And while I raise the cry, Help ! Help ! brethren, Help ! I wish to make more extensively known what has been successfully done in arresting the fatal effects of Malaria in Italy.

Of this Professor Drummond writes :—' The really appaling mortality of Europeans is a fact with which all who have any idea of casting in their lot with Africa should seriously reckon ; none but those who have been on the spot, or have followed closely the inner history of

African exploration and missionary work, can appreciate the gravity of the situation. The Malaria spares no man; the strong fall as the weak; no number of precautions can provide against it; no kind of care can do more than make attacks less frequent; no prediction can be made beforehand; as to which regions are haunted by it, and which are safe. It is not the least ghastly feature of this invisible plague, that the only known scientific test for it at present is a human life. That test has been applied in the Congo region already with a recklessness which the sober judgment can only characterise as criminal. It is a small matter that men should throw away their lives in hundreds, if need be, for a holy cause; but it is not a small matter that man after man, in long and in fatal succession, should seek to overleap what is plainly a barrier in nature, and science has a duty in pointing out that no devotion or enthusiasm can give any man a charmed life, and that those who work for the highest ends will best attain them in humble obedience to the common laws. Transcendently this may be denied; the warning finger may be despised, as the hand of the coward and the profane; but the fact remains, the fact of an awful chain of English graves stretching across Africa. This is not spoken nevertheless to discourage missionary enterprise, it is only said to regulate it.'

It is possibly the case that there are more missionaries, including under that designation all who have gone to central Africa at their own instigation or prompting, and it may have been at their own expense, and those who have literally been sent thither, together with their wives, who were not less missionaries than were they, inclusive of teachers, handicraftsmen and merchants, and their children, who have perished under the deadly influence of Malaria, than others emigrating thither from Europe; nor is a reason for this far to seek, when there is considered together with the proportional numbers in which they have gone forth there to settle for life, and their risk of life when once settled and engaged in their special

work. It has been alleged that sometimes they have
been imprudent in making choice of a site for their home.
Whatever ground may exist, or be supposed to exist, for
this opinion, it should be known that in some cases they
had no choice. As stated by Professor Drummond in
the passage just cited, there have been cases in which
there was no test which could be applied but the test of
experience, and this involves the alternative of life or
death. In regard to Livingstonia, the case first cited
above, he writes:—'I spent a day or two in the solemn
shadow of forty-one of that deserted manza. It is one of
the loveliest spots in the world; and it was hard to believe,
sitting under the Tamarind trees by the quiet lake shore,
that the pestilence which wasteth at midnight had made
this beautiful spot its home. A hundred and fifty miles
north on the same lake coast, the remnant of the missionaries have begun their task again, and there, slowly against
fearful odds, they are carrying on their work. Travellers
have been pleased to say unkind things of missionaries;
that they are sometimes right, I will not question, but I
will say of the Livingstonian missionaries and of the
Blantyre missionaries, and count it an honour to say it,
that they are brave, efficient, single-hearted men, who
need our sympathy more than we know, and are equally
above our criticism and our praise.'

I cannot bear testimony from personal observation,
but it is my belief that the same thing might be said with
truth of the other missionaries in the field, and of many,
besides the missionaries, who are there; and this I write
lest it should be imagined that they unnecessarily expose
themselves, either· in a reckless or a Quixotic spirit, to
unnecessary perils.

With what I know of the matter, I can affirm, that
apart from every higher consideration, the pecuniary loss to
an ecclesiastical or other organisation sustained through
the death of an agent of ordinary attainments sent out
from Europe or America to Africa to labour must be
much greater than need have been the expense of sur-

rounding the station with a broad belt of Blue gum-trees; and again, apart from the question of pecuniary profit or loss, justice requires that those who are sent forth to a malarious region to carry out the purpose of those sending them, whether it be to enable them to increase their gains and comforts, amenities and luxuries, while they and their families reside in a salubrious home, or it be to carry the gospel of salvation to the heathen, towards the expenses of which they are contributing of their abundance, should have every reasonable provision being made against unnecessary suffering or death in carrying out the enterprise; nor do I question that this would be done extensively if it were generally known how it might be done. It is under this impression that this Tractate has been published, and the more comprehensive volume mentioned in the foregoing circular has been prepared. In that there is given more detailed information in regard to the culture of the tree in Africa, Italy and in other countries of Europe, in Asia, in Australia, in America; information in regard to the natural history of the tree; information in regard to the natural history of the microbe, supposed to be the cause or occasion of the disease; and, information in regard to the disease itself. As stated in that circular, the compiler is ready to give the manuscript to any one who will meet the expense of printing it for extensive distribution in lands in which such fevers prevail, on the sole condition that fifty or a hundred copies be placed at his disposal.

III. Report by M. Auguste Vallée on the culture of the Eucalyptus at the Monastry of St. Paul, Tre-Fontanes, near Rome.

M. Auguste Vallée, Agricultural Engineer, Member of *La Societe des Agriculteurs de France*, of *La Societe D'Agriculture D'Indre-et-Loire*, &c., purposing to study agriculture in Upper Italy, while previously sojourneying for a time in Rome, visited the Monastry of Tre-Fontanes.

In regard to the name of the Monastry, he states that,
according to tradition, Paul suffered martyrdom on the
spot on which one of the Churches of the Monastry
stands, and tradition tells that when the head of the
Apostle was dissevered from the body, it rebounded three
times, and every time it touched the ground there sprung
up a fountain of water; hence was given the name *Tre-
Fontanes*.

There are three Churches within the Abbey, the
largest, that of S. Vincent and S. Anastase, is a Roman
basilique, built by Honorius I. in the Ninth Century. It
was restored in 1221, and is still in the state of its
restoration. On the right of this, the second Church,
octagonal in form, owes its name of Sante-Marie Scala-
Coeli to a vision of S. Bernard, to whom Innocent III.
had entrusted the charge of the Convent. It dates from
the Sixteenth Century. And the third Church, S. Paul
Aux Tre-Fontanes, encloses the fountains. It was built
in 1599.

Of the Trappists it may be told that they are a very
self-denying order of monks, originally founded in the
valley of La Trappe, from which their name is derived.

'Leaving Rome by the gate of S. Paul,' says M. Vallée,
'it is a journey of about five kilometers to the Abbey.
It appears tranquil, fresh and smiling in the open Roman
Campagna. On the left hand, before entering the
Monastry, the eye finds delight in the verdure of the
meadows, well managed and producing herbage of excellent quality. Beyond these meadows, forming the
basins of lovely little valleys, arise small hills covered
with fertile vineyards, and little woods of the blue gum-
tree. On the right hand, the view is in the distance
bounded by an unsightly uninterrupted succession of
narrow valleys and mounds, the bowels of which have
been deeply furrowed by the miners of *Pouzzolane*.

'Before 1868, the time at which the Trappists took
possession, this Monastry of ancient foundation had been
long abandoned. Nor is it difficult to discover a cause

for the abandonment. Situated in the centre of a country
where the terrible *Malaria* brings sore suffering and
sowes the seeds of death, the Romans had given to it
the designation, *The Tomb*, a name only too well merited.
Here the Trappists came at length to pitch their
tent. They were not ignorant of the blanks unceasingly
made by the Malaria in the population of the Agro
Romana, and they came to expose themselves to it, not as
men daring in stupidity, or unconscious through ignorance,
but with a brave heart and a strong will, which had a
noble end to accomplish. A noble end, indeed! and who
better fitted for its accomplishment than these men who
had no tie in life, who had abandoned country and home,
who once for all sacrificed their all, men strong and
energetic, who advanced to the duty calling them without
thought of returning! A noble end! it aimed at nothing
short of making this desolated region healthful, to make
this pestiferous land a nursing mother, and this by works
of improvement through the culture of a plant providenti-
ally brought just then to the country. During the first
years of the enterprise the fever spared not the useful
unostentatious lives. She found them in the midst of
their labours, and slew many victims. But this had no dis-
couraging effects. The fever raged, struck, and slew
pitilessly. The more the Trappists saw it face to face,
the more did they long to overcome it, and the more
resolutely did they hurl themselves determinately against
it. An enemy unknown occasions more fear than an evil
looked in the face. Their efforts, their energy, their
works, have already been largely recompensed, though the
last word has not yet been spoken. The fever has continuously lurked within the abode of the labourers, but
if sometimes it still puts in a furtive appearance, they
have arms with which to hunt it, and as it does not feel
itself the stronger, it must yield.'

There are detailed in the report the generic character-
istics of the Eucalyptus, and the author goes on to say:

'They have tried the culture of a great many species
of Eucalyptus at Tre-Fontanes, but they have only had

satisfactory success with the following: E. Globulus, E. Resinigera, E. Rostrata, E. Urnigera, E. Tereticornis, E. Cocceifera, E. Viminalis, E. Melliodora, Gunii, Stuartina and Red Gum.

'These Eculypti, moreover, do not all grow equally well on the same kind of ground. Thus, on moist ground it was found very preferable to raise only the following varieties: E. Viminalis, E. Urnigera, E. Rostrata, E. Tereticornis. On dry lands, on the contrary, they planted only these: E. Resinifera, E. Melliodora, E. Sideroxylon. The E. Globulus, or blue gum, grew well on almost all kinds of soil.

'At Tre Fontanes, as in almost all high land of the Agro Romana, the soil is of volcanic origin. The thickness of the layer of arable land is very variable, ranging between 20 and 40 centimeters. The composition of this layer varies with the different positions which it occupies. Thus, the ground on the summits of rising grounds has much less humus than that on the slope, and of course much less than that in the hollows.

'The predominating kinds of earth at the Fontane are these: clay land, sandy clay, alluvial land containing clay and sand, and embodying much more organic matter than the others.

'The soil certainly possesses sufficient fertility to warrant cultivation, but unhappily the subsoil does not admit of the culture of shrubs or other plants having a long tap-root. The subsoil is quite impermeable; it is a kind of lithoid tuffa, called by the people of the country cappellaccio, or the puzzolana cap.' Another writer on the subject, M. Meaume, states in illustration of the use made of the name, ' The Italian language is rich in diminutives augmentatives, thus: an ordinary hat is a cappello; and a small hat, cappellino; a large hat, cappellone; an old and useless bad hat, cappellaccio.' From this use made of the last term has it been applied to the volcanic product in question, the material is of no use but has to be broken up. Hard as it is it cannot be made use of in building nor in road-making. The puzzolana mixed with a fifth part of chalk makes the so called Roman cement, of excellent quality.

This layer is very variable in thickness: M. Vallée found it in different quarries, at the gate of the monastry, from 50 centimeters to 2 and 3 metres.

'Then comes the puzzolana proper, or volcanic sand, varying in colour with the position which it occupies. Grey immediately under the cappellaccio, with a thickness of from 50 centimeters to 1 and 2 metres, then red sand to a very great depth. It is this which is made use of in the manufacture of Roman cement, and no other sand is made use of in the buildings of Rome and in the Roman Campagna.

'The subsoil was the great hindrance to be overcome in the cultivation of the ground. The agricultural implements in use could do nothing with the subsoil. What then was to be done? It was impossible to establish a plantation of trees on a soil 30 centimeters thick, in these circumstances it occurred to the Trappists that the new force dynamite might be made use of. It was a happy thought, and led to important results !

'At first, to make the small mines necessary, the brethren made use of an iron bar with a steel bit, the only implement then known, which was struck with a mallet and slightly moved about by hand. But they soon found it necessary to devise some more expeditious and less toilsome means, and they contrived a hand borer which the Brothers Conscience, Mechanicians in Rome were employed to make.

'The disintegration of the tuffa is effected by a borer or auger of a particular form, to which is given a rotatory movement of from 3 to 4 turns in the second. This borer is fixed in a *manchon* or on the lower extremity of an iron axis along which is a groove. Two driving gears of conical form placed vertically and receiving movement by two cranks, transmit this to a third horizontal driving gear carrying in its axis a fixed key. This last gear receives

in its central hole the borer-bearing axis which is fixed
by the key. In this way the borer receives a rotatory
movement, but is free to mount or fall as may be
necessary. To effect the double movement there are two
racks connected with the two extremities by two pins
opposite to and commanded by two gears, which are
supported by a horizontal beam bearing a fly wheel and
crank. The whole is protected by an iron case, and the
whole is fixed on two supports fastened on a wooden
basis bearing swinging pivots, which mode of adjustment
admits of the machine being always in a horizontal
position. A solid wooden framework with four feet and
transverse beams sustains the whole apparatus. At the
extremity of each cross beam is an adjusting screw to
meet the inequality of the ground on which may be
placed the perforater. Movement being imparted to the
cog wheels, the point of the borer tears the rock, and
through the weight of its axis and racks, buries itself
automatically in the ground. As soon as the resistance
under the crank is felt to be increased, it is necessary to
bring into play the two pinions which act on the racks.
Then the borer retires quickly drawing out the debris torn
from the rock. The borer is then made again to descend,
without the cranks being stopped, until the hole has attained
the depth desired, from 0·80 to 1·20 metres. The time
occupied with a perforation is on an average from 5 to 6
minutes. The machine costs little; and the saving is great:
what it effects in 6 minutes a skilful miner could not do
with hand in less than 20 or 25 minutes. By this machine
the subsoil is broken up to a great depth; and the
mixture of soil and subsoil constitutes an earth of good
quality, possessing physical properties very favourable to
culture. The operation enables the Trappists to raise
also the level of the valleys, by hurling the earth from
the summits of the hills. The work has been tedious
and costly, but the results have proved most satisfactory.
The ground is thus prepared to receive the plants; but
the preparation of these must also receive our attention.

'The Trappists sow the seed of the Eucalyptus in

boxes or pots filled with earth prepared with some care. This earth is a mixture of the earth of the outlying field with earth from the garden, with a small quantity of well decomposed manure added to it; the whole well broken up and passed through a sieve is then placed in these receptacles. There is traced on the surface of this earth very shallow lines, distant from each other about six centimetres, and in the bottoms of these lines are deposited the seeds, which being extremely small are covered with a very thin layer of earth. The sowings may be made in February, and in March, and in Autumn.

'To retain the moisture needed, these boxes are then covered with a mat or a deal until the germination of the seed, which generally takes place in a few days, and from this time commences the particular care which has to be bestowed upon the plant for some months. It is necessary to prevent the young plants being chocked by the growth of weeds, to prevent too great desiccation of the earth, to see that the heat does not become too great for the plants, and to protect them from violent winds.

'When the plants are about 4 inches in height they are transplanted into flower pots about 8 inches in diameter at top and about the same in depth, and in these they remain till they have to be planted out in the ground; but they are also sometimes transplanted at first into boxes about 8 inches deep, and 70 centimetres or 28 inches long, which may receive 40 plants. It is requisite that these boxes be such as can be carried after the planters. Two men can easily carry them. This is preferable, especially in extensive plantations, in the open country. They are followed by a cask of water that each plant may be watered when put in the ground; and it is necessary to repeat the watering once-a-week during the first year of their growth.

'This planting out is done in spring. This season is preferred for the work at Tre-Fontanes because the young plant has thus the longer time to strengthen itself against the rigours of winter; but it may be done in September and October, in which case, in general, the autumnal

rains render unnecessary the hand watering. When the
plants are to be planted out in Spring, the sowing should
take place in the preceding Autumn, so that the plants
may be six or eight months old when planted out. The
work of planting is a very simple one, there are
first made long trenches 80 centimetres or 32 inches in
depth and of similar breadth, and therafter the earth
taken out is returned into the ditch. In this soft earth
there are dug holes 25 centimetres or 10 inches in depth
and of similar diameter; in this the plant is deposited, a
plant taken from the flower pot or box, with as much
earth around the root as possible, care being taken to
place the plant a little lower than the level of the soil so
as to leave a little cup or depression in the ground to
retain the water when the plant is watered or when it
rains. In the following year this is filled up to the level
of the soil. At Tre-Fontanes at first they planted the
trees at a distance of a metre or 40 inches from each
other in all directions; but after a years growth they
had to root out nearly one-half of them, because after a
year's vegitation they pressed upon and injured one
another. Now they plant the saplings two centimetres
or well nigh 7 feet distant from each other.

'The tree is planted, but man has not yet finished his
work; he cannot leave the young and frail sapling to the
care of nature yet; it is needful to protect it against
invasion by weeds. There must be frequent weedings
executed during the first two years at the foot of the
tree, and to a distance of from 20 to 24 inches; the air
and moisture are thus enabled more easily to reach the
rootlets, and vegitation goes on successfully.

'At the Monastry of Tre-Fontanes it cannot be said
that there are no winds nor violent winds; for such is
not the case; still it is very rarely that props are supplied to the young trees. At the end of a very short
time, and from the second year of growth, the plant has
become sufficiently vigorous not to suffer from the wind;
if this be an enemy to the Eucalyptus it is an enemy
which it can withstand; and it generally does so successfully.

'Another enemy of this plant is water when it is excessive in quantity, and in a state of complete stagnation. With this also it has to contend at Tre-Fontanes. There, there is water in abundance, and such water! Water stagnant and exhaling over the whole region deadly miasmata. By means of covered or open drains with sufficient slopes they have drained these off in such a way and direction that the ground as well as the atmosphere has been rendered healthy. Thus has another enemy been overcome.

'A last obstacle to the culture of the Eucalyptus there, and it the most formidable of all, is the cold. It is said that the Eucalyptus cannot live, in other words that it must die when the temperature falls to 6° or 8° centigrade below zero. It is true that this plant, the product of warm countries, has considerable dread of low temperatures, and although it can be shown that it has withstood temperature of 9° below zero, it must not be exposed to the open air in a country in which the temperature sinks repeatedly in succession to 5° or 6° below zero. In 1879 the winter here was very severe, but still all the plants above 2 years old withstood the cold, which was intense. The thermometer remained a long time at 9° below zero.

'But let us see to what extent, if any, the cold has interfered with the culture of the Eucalyptus in the Agro Romana. The Observatory of the Roman College gives 0.89 cent. as the mean winter temperature during a period of 24 years, from 1782 to 1861. The following are the minima observed during a period of 12 years from 1863 to 1874.

Years.	Minima.	Dates.
1863	4.64 centigrade	19th January.
1864	4.44	1st January.
1865	3.72	10th December.
1866	4.51	28th February.
1867	6.06	27th February.
1868	3.79	17th February.
1869	4.82	18th February.
1870	6.70	6th February.

Years.	Minima.	Date.
1871	5.55	19th January.
1872	3.71	2nd December.
1873	4.31	11th January.
1874	4.91	1st February.

'We have for the mean minimum of these twelve years 4.77; and never once has the temperature fallen below 6.70; and it is noteworthy that none of the other minima have occurred oftener than once. In 1875 the temperature fell to 9 centigrade at Tre-Fontanes. Happily this unusual and extraordinary cold was only experienced once, for in that one day almost the half of the plantation of the year perished. It is manifest that the Eucalyptus cannot sustain such a temperature. But this is an occurrence which has only happened once in a hundred years, and consequently should not be made a pretext for shackling this beneficial culture.

'It must in fine be told how, in despite of wind and of cold, the Eucalyptus have gone on growing, and what their wonderful increase has been, figures will speak more efficiently than any fine words, and they possess the advantage of expressing an indisputable fact. All the measurements which follow were made on the 8th April, 1879. In a plantation of *Eucalyptus Globulus*, formed in 1875, I have ascertained the following to be the measurements of a certain number of trees taken at random.

'Circumference taken at 1.50 metre from the ground in centimetres, 30, 20, 31, 28, 26, 30, 30, 30, 50, 20, 49, 28 23, 22, 28, 14, 15, 21, 27, 20, giving a mean of 27 centimetres.

'Heights in metres, 8.50, 7, 8, 6, 7, 11, 7, 10, 7, 6, 11 mean 8 metres.

'Measurements of E. Globulus planted in 1872.

	Circumference,	0.76 m.	Height, 13.50 metres.
		0.67	11.50
		0.62	11.50
		0.86	15
		0.87	17
Mean		0.75	13.50 metres.

'E. Globulus planted in 1870, circumference, 0.90 m.
Height, 15.50. E. Globulus, planted in 1870-1871, circumference, .78, height, 15 m.

'At the end of 1881 the trees planted in 1870, 1871, 1872, had a mean circumference of 1.15 metre.

'The following measurements were taken with great care at different times.

Species.	Year of Plantation.	Height in 1877.	CIRCUMFERENCE.		
			Near the Ground.	1 Metre.	2 Metres.
Globulus	1870	16 metres	0, m 95	0, m 81	0, m 75
„	„	16 „	0, 69	0, 62	0, 56
„	„	15 „	0, 79	0, 70	0, 61
„	„	16 „	0, 72	0, 60	0, 55
„	„	17 „	0, 90	0, 67	0, 62
„	1871	16 „	0, 76	0, 61	0, 55
„	1872	14, 50	0, 92	0, 75	0, 67
„	„	14 „	0, 90	0, 76	0, 66
„	„	16 „	0, 68	0, 58	0, 50
„	1873	15 „	0, 71	0, 50	0, 45
„	1872	15, 50	0, 85	0, 60	0, 49
„	1871	16 „	0, 86	0, 66	0, 60
„	1872	16 „	0, 90	0, 73	0, 70
„	1874	12, 50	0, 45	0, 42	0, 37
Viminalis	„	10 „	0, 45	0, 30	0, 27
Globulus	„	9 „	0, 36	0, 28	0, 24
„	„	8 „	0, 56	0, 42	0, 36
„	„ aug.	7 „	0, 47	0, 39	0, 31

'All the trees planted out in 1870 had at that time a height of from 15 to 20 m. One portion had been in pots; another in boxes. One of these plants having been measured near the ground after the first year, measured on the 6th March, 1874, 50 m. in circumference near the ground; and 35 at the height of 1 metre. On the 7th January following, that is after 10 months growth, it measured .62 near the ground, and .45 m. at the height of 1 metre, and it had a height of about 10 metres.

'With a statement of some other results which I have

observed in more recent plantations I shall close by statements in figures.

Species.	Year of Plantation	Height in 1877	CIRCUMFERENCE.			Obsrvtns.
			Near the ground.	1 metre.	2 metres.	
Globulus	Aug. 1874	9 metres	0, m 56	0, m 42	0, m 37	*Plants 2 metres apart in every direction in ground, trenched to the depth of 1 metre or 40 inches. The plantations left almost entirely to themselves.*
,,	,,	8 ,,	0, 51	0, 38	0, 32	
,,	,,	7 ,,	0, 49	0, 38	0, 33	
,,	Mar. 1875	6 ,,	0, 38	0, 295	0, 27	
,,	,,	8 ,,	0, 50	0, 40	0, 33	
,,	,,	8 ,,	0, 49	0, 40	0, 35	
,,	,,	7 ,,	0, 40	0, 29	0, 23	
Resinifera	Apr. 1876	4,50 ,,	0, 35	0, 18		
,,	,,	4 ,,	0, 28	0, 15		
,,	,,	4,50 ,,	0, 325	0, 22		

'Where shall we find in our forests a tree presenting a
spectacle of vegitation so marvellous? What other kind
of tree can under our climate attain at the end of eight
years a circumference of .90 m., and a height of 16 metres,
54 feet.

'With regard to the physical properties of the leaves
and of the wood, it is stated that certain varieties of the
Eucalyptus absorb from the ground a quantity of water
which is truly surprising. In the case of the Globulus a
square metre of foliaceous surface can pass into the atmosphere at least 2,400 kilogrammes of water. I say *at
least*, for the day on which I made the experiment, and in
those following the transpiration was not so active as
usual in consequence of the feeble intensity of the heat.

E. Resinifera has evaporated 3 kilogrammes of water
in 12 hours, from a square metre of foliaceous surface.
E. Rostrata from the Isle de la Reunion has evaporated
1.450 kg. from the same extent of surface in the same
time.

Now a leaf of E. Globulus having 90 centimetres of
square superfices, two faces being comprised in this
measurement, as from both evaporation takes place,

weighs 0.0115 kg., so that a square metre of foliaceous surface weighs 4.276 kg., according to the following formula.

$$4 \text{ m } 2 = 0.0115 \text{ kg.} \times \frac{10,000}{90} = 1.276 \text{ kg.}$$

'From all which it follows that the weight of water evaporated is about double or treple the weight of the leaves. The evaporation going on in the open air and under the influence of the wind being still more considerable, one may say, without fear of being guilty of exaggeration, that the plant evaporates a quantity of water at least four or five times the weight of its leaves. The respiration will also be naturally supposed to be very active if account be taken of the number of respiring organs with which the leaves are everywhere furnished. I have been able to count on a square millemetre of the inner surface of a leaf of the E. Globulus 350 stomates. Is it then astonishing that these plants can produce such an effect upon the air as has been referred to?

'From the Euclayptus there are several economic products procurable besides those particular essence which are obtained from the leaves: there is obtainable from this tree a gum-resin, known in commerce as Gomme de Chine. This product, which becomes red in solidifying, is obtained by means of incisions made into the bark. According to weight, a single tree may yield about 60 litres of this gum. Having seen with what rapidity both the growth and increase of this tree goes on, it may be supposed, in view of this, that the lignous tissues of the plant must be gorged with sap, and that the wood must be soft. It is not so; but very far from it. 'A completely dry cutting of the wood of the E. Globulus, taken the whole length of the diameter of the trunk, gave me as the specific weight the figure 0.836. This cutting was taken from a tree 8 years of age. Now the heaviest oak, Q. Ilex, weighs 1.083; the Q. Cerris weighs only 0.753, and the mean specific gravity of the different oaks growing in Italy is 0.852. Consequently, the wood of the Eucalyptus must be almost as heavy as the greater part of the different kinds of oak wood. Moreover, this wood of the Eucalyptus is considered so hard and solid that it is already employed in naval structures.

'I say nothing of the properties more or less marvellous which have been attributed to this tree; it belongs to chemists and medical men to speak of these. What I wish especially to give prominence to is the fact that the Eucalyptus is a tree particularly adapted to render salubrious the air, because of its rapid vegetation and its active respiration. But I would add that this property is not peculiar to the Eucalyptus, and that all vegetables, those with caducous leaves especially, possess also in a high degree the power of purifying the air under the influence of light. One can well understand that this may be one reason why the Trappists at Tre-Fontanes cultivate their other vegetables besides the Eucalyptus; for example, the vine: and from this I may take occasion to make some remarks on their vineyard. On the 32 hectares, of which the lands of the Monastry consist, there are 10½ planted with vines; and of these, 6½ are already in full bearing; the four others have been planted 1, 2, or 3 years only.

'On those lands which are of volcanic origin, and the depth of which has been considerably increased by tillage, the vine succeeds to a wonder. They cultivate several varieties of vine, which all yield what would be excellent wine if it were only a little richer in alcohol: for its weakness in this respect is an obstacle to its good preservation. The varieties which the Trappists cultivate are: the *Grenache*, to which they give by preference a south and south-east exposure. It yields 60 hectolitres per hectare. The *Carignan*, the vegetation of which in spring comes on a little more slowly. It prefers drained land; its yield is also 60 hectolitres per hectare. The *Espar*, which has a vegetation still more tardy; but it ripens its fruit at the same time as the others; it prefers poor soil. The *Clairette* is also cultivated, but it does

not always give satisfaction at Tre-Fontanes. It yields
little; but its wine is very fine. The *Aleatico* succeeds
very well; it likes very fertile land and a southern
exposure. It produces about a litre per vine stock. The
Trebiano is cultivated to the same extent; but it does not
accommodate itself to the short cutting which is in use at
the Monastry.

'Looking away from the question of profit, to look
only at that of amelioration of soil and climate, is it not
manifest that quite as well as the Eucalyptus the
enormous leaf production of the vine must be capable of
absorbing the fatal miasmata of the Agro Romana?
And, happy coincidence, it is precisely at the moment
when the Malaria begins to make itself felt that the vine
begins to cover the earth with its thick foliage.'

The conclusion drawn by M. Vallée in view of these
facts is that it is to be wished that other attempts at
amelioration should be made in the Agro Romano. He
expresses his belief that proprietors of this desolate
land would in no way injure their own interests by doing
so, while they would be doing a work eminently humane.
He says 'I know what the reply will be, it is that the
Agro Romano yields to proprietors 5 and 6 per cent. in
providing for their flocks an abundant pasturage; and are
they going to sacrifice this high and certain revenue for
an enterprise which will require of them an enormous
capital, force them to expose to the malaria a great
number of workmen, and all for a result which does not
seem to them to be by any means certain? Those who
speak thus would have some reason on their side if what
was advocated were the general and immediate transfor-
mation of the Agro Romana. But might not this work
of amelioration be done in small portions without as a
necessary consequence depriving the proprietary of capital
required for their maintenance? And as for pecuniary
returns, it is quite certain that in the places in which
the vine could be cultivated, the returns would rapidly
rise to more than 6 per cent. Besides, where were planted

the Eucalyptus or any other kind of tree, the return
though not immediately convertible into money would be
not less certain ; in any case one thing in no way doubt-
ful is that the property will rise annually in value in
proportion to the works executed upon it.'

In an Appendix to one edition of the Report by M.
Vallée, it is stated that in October, 1879, the Italian
Government made a further concession to the Trappist
brotherhood to enable them to carry out their important
enterprise on a greatly extended scale ; and subsequent
reports show that while previously the Friars could not
sleep without risk of life in the Monastry, but had every
evening to betake themselves to a Convent within the
Imperial city to sleep, returning in the morning to Tre-
Fontanes for the discharge of their ecclesiastical functions,
now they sleep there constantly ; and the peasantry who
had previously to betake themselves with their families
and all their portable possessions to the mountains in
the beginning of summer, whence they returned with
their goats and other possessions in the late autumn, now
reside on the Campagna 'from June to January,' and
from one end of the year to the other.

And while the Trappist Friars have been thus ameliorat-
ing the sanitary condition of the Campagna, statesmen and
others have, with similar effect, had the Blue gum-tree
planted in many other malarious districts in Italy ; and
similar effects have followed the culture of the tree here
and there and everywhere throughout the land. Since
then, there has been discovered a microbe 'like to
others to which have been traced the introduction of
many deadly diseases into districts where, at the time,
they were non-existent ; the natural history of this
microbe has been studied ; and, it has been designated
Bacillus Malariae. From this discovery there may
result new treatments of the disease ; but to prevent is
better than to cure ; and nothing has occurred to shake
confidence in inferences which may be legitimately drawn
from the sanitary effects of the culture of the Blue gum-
tree in Italy.

Lately Published, Price 12s

MANAGEMENT OF CROWN FORESTS

AT THE

CAPE OF GOOD HOPE

UNDER THE OLD REGIME AND UNDER THE NEW.

Compiled by JOHN CROUMBIE BROWN, LL.D.

EDINBURGH : OLIVER AND BOYD, TWEEDDALE COURT.

LONDON : SIMPKIN, MARSHALL, & CO.

1 8 8 7.

EXTRACT FROM PREFACE.

Part I. shows a waste of forest produce having gone
on at the Cape under the old regime greatly exceeding in
money value the amount of the revenue received from the
forests, and that the revenue was not great, being in 1862
about £250. While Part II. contains statements by the
superintendent of the forests to the effect that in the year
1882, after two years' management under the new regime,
the gross revenue was £7,680 14s ; and that the forests of
the Colony, if properly managed, might yield a revenue of
at least £235,000.

And this enormous increase of revenue is only one of several co-related benefits obtained under the new regime, some of which are only indicated, not detailed, in the reports cited.

Of the extent to which parsimony, apart from all idea of recklessness or imprudence, may account for the management of the same forests under the old regime, illustrations are supplied in Part III.

The details given are copious. They are so because I design my Treatise to be, in some respects, like to a clinical lecture, teaching from a particular case, supplying at once well defined indications of disease, explicit statements in regard to the remedial measures followed, official statements in regard to results obtained, and facilities for indicating the cause or occasion of the evil remedied.

I have prepared a similar account of Sylviculture at the Cape of Good Hope under the old regime and under the new. It is equally with this illustrative of the advantages which have thus been secured, declarative of what has been done there, and suggestive of what might be done elsewhere.

I have arranged that should any British Colony, or any American State, or any private individual, desire 50 copies for distribution, they may be supplied with these at a reduction of 25 per cent., or one-fourth of the retail price, or 100 copies at a reduction of $33\frac{1}{3}$ per cent., or one-third of that price.

JOHN C. BROWN.

HADDINGTON, 23rd June, 1888.

NOTICE IN THE SPANISH *Revista Contemporanea*, 15th AUGUST, 1887.—TRANSLATION.—" The untiring activity of Dr. Brown astonishes us exceedingly. We have scarcely finished reading one of his excellent works, when another appears. We are certain that a Forestal Library of the first order might be formed with the books published, within a few years past, by the learned Professor. Dr. Brown has described for us the forests of Germany and Spain, those of England from the most remote times, those of Finland, and of Northern Russia. He has dedicated a volume to the examination of the celebrated French Ordinance of 1669, and to the Plantations on the Lands of Bordeaux, and another to the Reboisement carried out in France ; there is one devoted to a most exhaustive Treatise on the Hydrology of South Africa, one not less interesting is devoted to the influence exercised by Forests on Humidity of Climate, and this, besides others which we do not at the present moment remember. And now this illustrious Scottish writer, whose 79 years of age do not weigh him down, surprises us with a conscientiously thought out work entitled *Management of Crown Forests at the Cape of Good Hope under the Old Regime and under the New.*

" Everything in regard to the forests owned by the Crown at the Cape is investigated by Dr. Brown with the competency given by his great learning, and his long residence at the Capetown, where he was Professor of Botany, and whence he made numerous lengthened excursions. This last work of his is arranged in three parts,—in the first is given an exposition of the treatment of forests there under the old regime ; in the second is given an explanation of the modifications introduced under the new regime ; and in the third is a comparative study of the differences observable in the forests under the one regime and under the other.

" The well-known works of Dr. Brown are so estimated amongst all Students of Forestry in Europe, that speaking of our own government we know that they would have had special pleasure in awarding to him one of the highest of our decorations, if the English law had admitted of this being done. As this is impossible, we hope Dr. Brown will accept our most hearty congratulations on the admirable activity with which he is promulgating the principles of Forest Science from his retreat at Haddington."

www.ingramcontent.com/pod-product-compliance
Lightning Source LLC
Chambersburg PA
CBHW031800090426
42739CB00008B/1090